T0401199

PINE CONE REGROWN

How One Species Thrives After Fire

ELISA BOXER

ILLUSTRATED BY KEVIN & KRISTEN HOWDESHELL

PUBLISHED *by* SLEEPING BEAR PRESS™

Nestled in the needles,
seeds sealed inside,
the pine cone
hangs high out of harm's way,
just as it has
for forty years.

Perched.
Protected.
Peering down
on the forest friends below:

Nesting.

Nibbling.

Gathering.
Grazing.

Far across the forest,
a forgotten campfire,
once crackling,
still smoldering,
sends up a spark.
Lands on a leaf.
Sticks to the bark.

Flames flicker.

Nestled in the needles,
seeds sealed inside,
the pine cone
hangs high out of harm's way.

Peering down
on the forest friends below:

Flying.

Fleeing.

Scampering.

Burrowing.

Giant fingers
of fast-moving fire
flying through the forest,
reaching for the sky.

Orange ghosts,
gathering speed,
haunting the trees.

For forty years,
the pine cone
has stood sentry.
Seeds inside
safe and snug and sealed tight
by resin—
a sticky substance stronger than glue.

Perched.
Protected.
Peering down
on the now empty spaces
where forest friends below
once

Nested.

Nibbled.

Gathered.

Grazed.

Just out of reach
of the raging flames,
the pine cone keeps its seeds safe
from the heat
for as long as it can.

But the fire starts climbing.
The bark starts burning.
The needles start dropping.
The resin starts melting.

The cone cracks open and the

seeds

start

falling.

They land in a layer
of ash
on the charred forest floor.
A scorched spot
that hasn't seen the sun
in forty years.

Until now.

With the canopy cleared,
the light
shines down.
Nutrients from the ash
act as fertilizer.

Next to the charred remains
of its parent pine,
the seeds,
set free,
start to sprout.

It wasn't the end.

From the fire,
life
unfolds.
Green
and new
and ready
to begin.

AUTHOR'S NOTE

Without a doubt, wildfires are destructive. They come at a tremendous human cost, including health impacts from smoke, loss of property, and loss of life. The higher temperatures and increased drought that come with climate change have made the problem worse. In fact, the book I set out to write was about how a forest ecosystem regenerates after the devastation of fire. In the course of my research, I came across the **lodgepole pine**, a so-called serotinous tree whose cones only open after a fire.

I couldn't stop thinking about them—these cones that need fire to create new life, this force of destruction as a simultaneous source of propagation.

The glue-like resin seals the seeds inside the cone. The scorching temperatures melt the resin. The cone cracks open, releasing the seeds down to the forest floor, where the soot and the sun nourish them to grow. New seedlings sprout up almost immediately, with the help of parent pines that die and fall, providing nutrients for this new generation of trees.

IT'S A DELICATE BALANCE BETWEEN DEVASTATION AND GROWTH.

It's a balance mirrored by the environmental necessity of fire, as well. Fires have always been a natural and essential element of a healthy forest ecosystem. Fires return crucial nutrients to the soil, for example, and prevent the overcrowding of trees that would lead to more severe destruction from a wildfire. For these reasons, controlled burns are sometimes used as a tool for forest conservation and regeneration. And in some regions, called fire-maintained ecosystems, fire is necessary for certain species to survive.

The lodgepole pine is one of those species, and one of several species of serotinous trees whose cones need fire to thrive. Others include the **cypress pine**, the **jack pine**, and the **table mountain pine**. Most are found in national parks throughout the western United States such as Crater Lake, Glacier, Rocky Mountain, and Yellowstone.

In many cultures, fire is seen as a source of transformation and transmutation. It burns away what was ready to go and clears the way for what's waiting to emerge.

Through adversity, we find our strength. And maybe, just maybe, like the serotinous pine cone, in challenging times we might find we were made for that very moment.

OTHER LIVING THINGS THAT REQUIRE FIRE TO THRIVE

The **AUSTRALIAN GRASS TREE** produces long blooming spikes after a fire. In fact, these trees are often found in greenhouses, where workers use blowtorches to promote flowering!

The **SOUTH AFRICAN FIRE LILY** will only flower after being exposed to the smoke from a fire. And it blooms quickly—often in a little more than a week after a fire.

The **BURN MOREL MUSHROOM** is among several species of fungi that can only bloom when exposed to the heat from a fire.

The **WOOD WASP** will often lay its eggs in trees burned by fire.

The **AUSTRALIAN FIREHAWK** hangs out in burning areas to find food. When insects, lizards, small birds, and other animals start running away, the firehawks swoop in and grab them.

The **BARK BEETLE** makes its home in the blackened bark of a tree killed by fire. But look out, bark beetle! A black-backed woodpecker looks for these dead trees, knowing there are tasty beetles inside.

SELECTED SOURCES:

BOOKS

Furgang, Kathy. *Wildfires*. Washington, DC: National Geographic Partners. 2015.

Hinshaw Patent, Dorothy. *Fire: Friend or Foe*. New York, NY: Clarion Books. 1998.

Ohlsen, Erik. *The Forest of Fire*. Sebastopol, CA: StoryScapes. 2016.

Peluso, Beth A. *The Charcoal Forest: How Fire Helps Animals and Plants*. Missoula, MT: Mountain Press Publishing Company. 2007.

ONLINE

Green, Ruben. "Serotinous Survival Strategies." Evergreen Arborist Consultants, October 6, 2014.

Mullen, Luba. "How Trees Survive and Thrive After a Fire." National Forest Foundation, 2017.

Nix, Steve. "Serotiny and the Serotinous Cone." Treehugger, January 21, 2020.

Petruzzello, Melissa. "Playing with Wildfire: 5 Amazing Adaptations of Pyrophytic Plants." Britannica Online.

For Billy,
and the new life that could only have grown from the soil of the struggle.
—Elisa

To Matt & Claire,
we are so proud of the new growth we've seen in you.
—Kevin & Kristen

SLEEPING BEAR PRESS™
2395 South Huron Parkway, Suite 200, Ann Arbor, MI 48104
www.sleepingbearpress.com © Sleeping Bear Press

Printed and bound in South Korea
10 9 8 7 6 5 4 3 2 1

Library of Congress Cataloging-in-Publication Data on File.
ISBN: 978-1-53411-296-4

Photos: Lodgepole Pine: Rachel Portwood | Shutterstock.com • Forest Fire: AdventurousMama88 | Shutterstock.com
Australian Grass Tree: Darkydoors | Shutterstock.com • Burn Morel Mushroom: ressormat | Shutterstock.com
Bark Beetle: Nikolas_profoto | Shutterstock.com